ECOSYSTEM ALARM MANAGEMENT

MIT License

Copyright (c) 2023 Aidan Thomas Parkinson

Permission is hereby granted, free of charge, to any person obtaining a copy of this software and associated documentation files (the "Software"), to deal in the Software without restriction, including without limitation the rights to use, copy, modify, merge, publish, distribute, sublicense, and/or sell copies of the Software, and to permit persons to whom the Software is furnished to do so, subject to the following conditions:
The above copyright notice and this permission notice shall be included in all copies or substantial portions of the Software.
THE SOFTWARE IS PROVIDED "AS IS", WITHOUT WARRANTY OF ANY KIND, EXPRESS OR IMPLIED, INCLUDING BUT NOT LIMITED TO THE WARRANTIES OF MERCHANTABILITY, FITNESS FOR A PARTICULAR PURPOSE AND NONINFRINGEMENT. IN NO EVENT SHALL THE AUTHORS OR COPYRIGHT HOLDERS BE LIABLE FOR ANY CLAIM, DAMAGES OR OTHER LIABILITY, WHETHER IN AN ACTION OF CONTRACT, TORT OR OTHERWISE, ARISING FROM, OUT OF OR IN CONNECTION WITH THE SOFTWARE OR THE USE OR OTHER DEALINGS IN THE SOFTWARE.

First Edition 2023 (with minor corrections 2024)
ISBN: 978-1-7394554-1-5 (hardback)
https://github.com/aidan-parkinson/digital-assurance

Fringe Papers is a registered trademark of Realfeed Ltd.

Ecosystem Alarm Management

A Manifesto in Systems Architecture

Aidan T. Parkinson

Fringe Papers
Wedmore, UK.

For those passionate activists and disruptive technologists.

Abstract

It has been recognised that the prevailing discussion on the governance of platform ecosystems may have contributed to global ecosystem instability in alarming ways. The prioritisation of network effects to deliver a popular market share of consumers tends to be a common Feature. This investigation seeks to diagnose causes of instability and define new platform ecosystem boundaries for more stable outcomes. This has been achieved through an investigation of ethics, continuous integration processes and operations. This investigation appreciates the differences between the *states* of Locke and *State* of Hobbes. The Sovereignty of states and their role in forming civil society are recognised and this platform submits to the necessary assurance regime of a state. However, a significant difference of this platform ecosystem to others is the acknowledgement that a society need also submit to a global Sovereign of State which accounts for the condition of the Earth's ecosystem. What follows is the use of a Hobbesian Commonwealth Cost of Carbon as an overall quality of Life indicator and a Rawlsian process of alarm management. The resulting *Commonwealth of Peoples* could adopt traditional principles of justice amongst free and democratic Peoples without conflict. A minimal set of criteria has been identified for a Publicly addressable alarm management system that may allow complimentors to develop new Features for end-consumers. Evaluations of practice have identified a prioritised agenda of platform alarms with universal relevance, each of which have associated recommendations for action. In operation the platform is to share knowledge in concert with other Professionals to develop mutually assured approaches in Good faith.

Abbreviations

API	Application Programming Interface
ARP	Address Resolution Protocol
CCGT	Combined Cycle Gas Turbine
DAQ	Device Automated Qualification
DPCV	Differential Pressure Control Valve
FTP	File Transfer Protocol
GPL	General Public License
HTTP	Hypertext Transfer Protocol
IP	Internet Protocol
LTS	Long Term Support
MQTT	Message Queued Telemetry Transport
NAT	Network Address Translation
PAT	Port Address Translation
PITRV	Pressure Independent Thermostatic Radiator Valve
RADIUS	Remote Authentication Dial-In User Service
RDP	Remote Desktop Protocol
SCADA	Supervisory Control and Data Acquisition
SDN	Software Defined Networks
SSH	Secure Shell
TCP	Transmission Control Protocol
TRV	Thermostatic Radiator Valve
WAN	Wide Area Network

Contents

1 Epigraph   1

2 Introduction   3

3 Ethics   7
- 3.1 State of Nature 8
- 3.2 Sovereignty 13
- 3.3 Quality of Life 13
- 3.4 Minimum Decency 27
- 3.5 Rate of Saving 29
- 3.6 A Note on Perfectionism 32
- 3.7 Demonstration, Assurance and Automation   33

4 Continuous Integration   35
- 4.1 Life 38
- 4.2 Public 39

CONTENTS

4.3	Property	49
4.4	Alarm	49
4.5	Sensitive	51
4.6	Personal	51
4.7	Client	52

5 Operation   **53**
5.1	Agenda	54
5.2	Association	71
5.3	Licensing	72
5.4	Contracts	73

6 Conclusion   **75**

CHAPTER 1

Epigraph

"And now, the end is here
And so I face, the final curtain
My friend, I'll say it clear
I'll state my case, of which I'm certain
I've lived, a life that's full
I travelled each and every highway
And more, much more than this
I did it my way"

CHAPTER 1. EPIGRAPH

"Regrets, I've had a few
But then again, too few to mention
I did, what I had to do
I saw it through, without exemption
I planned, each charted course
Each careful step, along the byway
And more, much more than this
I did it my way"

"Yes, there were times, I'm sure you knew
When I bit off more than I could chew
But through it all, when there was doubt
I ate it up, and spit it out
I faced it all, and I stood tall
And did it my way"

"I've loved, I've laughed and cried
I've had my fill, my share of losing
And now, as tears subside
I find it all, so amusing
To think, I did all that
And may I say, not in a shy way
Oh, no, oh, no, not me
I did it my way"

"For what is a man, what has he got?
If not himself, then he has naught
To say the things, he truly feels
And not the words, of one who kneels
The record shows, I took the blows
And did it my way"

"Yes, it was my way" [1]

CHAPTER 2

Introduction

Digital platform governance involves thoughtful considerations towards the standardisation of interfaces to enable a diverse market of complimentors to find opportunity in developing Features for consumers. There has been much discussion about the creation of platforms that prioritise network effects to deliver a popular market share of consumers. Any platforms popular market share is often considered important in creating compelling development opportunities for complimentors to offer proprietary tools to enhance personal utility. Such an approach has been very successful for exploiting new domains, such as the internet and personal client devices [2]. However, such contributions may be somewhat detrimental in addressing societies pressing contractual problems involved in minimum security Standards to govern a limited natural ecosystem. Loss of biodiversity, the coercive politics of limiting global temperature rise, social unrest and military intervention are all relevant demonstrations in this domain and appear to exhibit some common Features.

One potential area of conflict in ecosystem governance is the management of competing principles of what is Good. This involves a definition that not only justifies ecosystem ethical priorities, but also a recognition of what is Sovereign. Minimal policies could then be identified to contribute towards ecosystem stability and align supply-chains with a natural course of development.

Another important aspect is to set boundaries for the development of distinctive attributes known as *Features*. Frameworks and Guidance need identifying that signal the failure of Features under development within an environment that supports continuous integration. Failure is an important feedback mechanism for developers to learn and adapt work-in-progress Features and, hence, enforcement should ideally be rapid, consistent and relevant. It has been noted that in practice recognised standards are poorly enforced for the networking of outstations and sensors in the field over Public domains. Lack of consumer confidence has contributed to sluggish market penetration of Internet of Things technologies to-date.

Further, it appears important to consider the management of Property and how governing agents may operate. Such thought needs to reflect on the ecosystems ethical priorities, competition and operational costs. In operation, a prioritised agenda of alarms with recommendations for action shall be itemised

Therefore, this paper asks the question:

How could one govern an alarm management platform to support a stable and diverse natural ecosystem?

The discussion draws upon experience of practice and a wide-ranging review of theory. In conclusion an Ontology that offers the authors perspective on this question is proposed. It is acknowledged that the problems faced are formidable and no solution can be perfect. However, it is intended that this paper contributes common-sense that may inspire the sustainable development of new technologies going forward.

CHAPTER 3

Ethics

Taking weighty responsibility, as this paper does through asking its opening question, a critical ethics statement must form the core of any platform governance. The issue being explored has relevance to the stability of the social contract and perspectives on the State of nature itself. What follows is a summary discussion of an ethical basis to be committed to by an assurance regime. It is believed that this wide-ranging perspective constitutes a comprehensive philosophical doctrine for development.

3.1 State of Nature

It is acknowledged that any ecosystem may include agents with different interests and potentially competing principles of what is *good*. However, it is extremely challenging to identify a global perspective of what a Good interest is. In an attempt to govern a natural ecosystem, one could understand that a process of identification involves considering prevailing representations of the State of Nature. Two influential accounts define significantly different interpretations of nature and association. These are John Locke Second Treatise and Thomas Hobbes' Leviathan. This document carefully draws distinctions between these two perspectives by use of the terms *states* and *State*

John Locke defines a *state* of nature for individuals [3].

> "...a state of perfect freedom to order their actions and dispose of their possessions and persons as they think fit, within the bounds of the law of nature, without asking leave or dependency on the will of any other man."

Locke then defines bounds of nature [3].

> "...no one ought to harm another life, health, liberty or possessions."

From Locke's perspective, some persons may transgress these bounds and in response people may defend themselves or others against such invaders of rights. The injured party and his agents may then recover from the offender [3].

> "...so much as may make satisfaction for the harm suffered"...

CHAPTER 3. ETHICS

> "...everyone has a right to punish the transgressors of that law to such a degree as may hinder violation."

Nozick describes how such a view leads to the development of protective associations to enforce what an individual believes to be their rights, with others joining a call for defence. To avoid mutually harmful conflict between protective associations, a system of appeal courts and judgments are made on jurisdiction and the conflict of laws in favour of the dominant protective association. Protection and enforcement of people's rights is treated as an economic Good to be provided by the market, as are other important Goods such as food and clothing [4].

Alternatively, Hobbes defines a *State* of nature for civil society. Hobbes assumes all people as naturally equal in body and mind. This includes an equality of hope in the attaining of ends. This leads to three principle causes of quarrel: competition for personal gain; diffidence for safety; and glory for reputation. Without a common power to stipulate rules a war of all against all ensues and a potentially fraught and divisive struggle [5].

> "NATURE had made men so equal, in the faculties of the body and mind..."

> "Hereby it is manifest, that during time men live without a common power to keep them all in awe, they are in that condition which is called war; and such a war, as is of every man, against every man. For WAR, consisteth not in battle only, or the act of fighting; but in a tract

of time, wherein the will to contend by battle is sufficiently known: and therefore the condition of time, is to be considered in the nature of war, as it is in the nature of weather. For as the nature of foul weather, lieth not in a shower or two of rain; but in an inclination thereto of many days together: so the nature of war, consisteth not in actual fighting, but in the known disposition thereto, during all the time there is no assurance to the contrary. All other time is PEACE."

"To this war of every man against every man, this also is consequent; that nothing can be unjust. The notions of right and wrong, justice and injustice have there no place. Where there is no common power, there is no law, no injustice. Force, and fraud, are in war the two cardinal virtues. Justice, and injustice are none of the faculties neither of the body, nor mind."

However, Hobbes asserts that there are certain conditions where men are inclined to peace: fear of death; desire for a certain standard of living; and hope for industry [5].

"The passions that incline men to peace, are fear of death; desire of such things as are necessary to commodious living; and a hope by their industry to obtain them. And reason suggesteth convenient articles of peace, upon which men may be drawn to agreement. These articles, are they, otherwise are called the Laws of Nature..."

CHAPTER 3. ETHICS

When people mutually covenant each to the others to obey a common authority, they have established what Hobbes calls *Sovereignty by institution*. When, threatened by a conqueror, they covenant for protection by promising obedience, they have established *Sovereignty by acquisition*. These are equally legitimate ways of establishing Sovereignty. The social covenant involves both the renunciation or transfer of right and the authorisation of the Sovereign power. Political legitimacy depends not on how a government came to power, but only on whether it can effectively protect those who have consented to obey it; political obligation ends when protection ceases.

Through transferring one's personal freedoms to a trusted Sovereign one sets up the conditions for *keeping of promise* and, hence, the conditions of stable currency. Such a Sovereign requires resources for enforcement to ensure performance of promises made and defend against competing externalities [5].

> "The mutual transferring of right, is that which men call CONTRACT."

> "... one of the contractors, may deliver the thing contracted for on his part, and leave the other to perform his part at some determinate time after, and in the mean time be trusted: and then the contract on his part, is called PACT, or COVENANT: or both parts may contract now, to perform hereafter: in which cases, he that is to perform in time to come, being trusted, his performance is called *keeping of promise*, or faith..."

> "He that transferreth any right, transferreth the means of enjoying it, as far as leith in his power. As he that selleth land, is understood to transfer the herbage, and whatsoever grows upon it; nor can he that sells. mill turn away the stream that drives it. And they that give to a man the right of government in Sovereignty, are understood to give them the right of levying money to maintain soldiers; and of appointing magistrates for the administration of justice."

Hobbes prescribes peace, promise and rule of laws as a Good for social benefit as far as there is hope to obtain it [5].

> "... that every man, ought to endeavour peace, as far as he has hope of obtaining it; and when he cannot obtain it, that he may seek, and use, all helps, and advantages of war."

Through taking the Hobbesian perspective into account, we may be led to the contributions of John Rawls as a substantive governance system for managing competing principles of the Good.

Through an exploration of conflicts between these two perspectives on the State of Nature, a proposition for an alarm managers executive function may be defined as follows.

3.2 Sovereignty

Following our exploration of a State of Nature, we understand that identification of a Sovereign is a necessary requirement of any ecosystem to be governed. The initial priority here is to address root cause of *uncaught promises* which may result from conflicts between the prevailing perspectives of Locke and Hobbes. An essential reliance on states to underwrite and re-negotiate the competing priorities of contractual obligations compromised through unexpected natural disasters entirely undermines justification of popular governance models grounded in Locke's state of nature. Locke's bounds of nature can only conceivably apply to civil society and entirely miss the potential of the Wild to bring anyone to harm. It is unrealistic to claim freedom from having to submit to a common natural ecosystem. A society of peoples needs to submit to a Hobbesian perspective of a Sovereign ecosystem for all humanity called *Earth*. The resulting *Commonwealth of Peoples* could adopt traditional principles of justice amongst free and democratic peoples without conflict [6].

3.3 Quality of Life

Understanding what constitutes quality of Life is a somewhat existential exploration for the human race. Today, we find leaders defining quality of Life as anything from household income to curated human development indices. This exploration has focussed specifically on what quality of Life means to the Peoples of the Global Commons. It is intended as an effort fit and equitable for all society.

Stewardship of the Global Commons

Carbon is a name for a non-metallic element with atomic number 6 and symbol C. On Earth this element has a key role in a *carbon cycle* that is essential for all forms of Life on the planet. This cycle involves exchanges of carbon-based vapours with the Earth's atmosphere. Plants sequester Carbon Dioxide (CO_2) through photosynthesis. Animals and plants emit CO_2 and animals Methane (CH_4) through respiration and metabolic processes. Industrial processes can also emit CO_2 and other carbon-based vapours to deliver Goods and services to support a growing human populations well-being [7].

There is evidence of a governance problem with the carbon cycle on Earth from the British Antarctic Survey. The British Antarctic Survey has studied ice cores and the air bubbles trapped within them from Antarctica and Greenland since the 1950's. Such research seeks to understand historic temperature and concentrations of molecules in the Earth's atmosphere. The studies provide evidence of stable concentrations of CO_2 since year 1000AD, until the onset of the industrial era in the early 19th century. Since then, concentrations have risen and were reported in 2010 to be approximately 1.4 times higher than before industrialisation. Such high concentrations of CO_2 in the atmosphere are thought to be unprecedented in the previous 800,000 years. Furthermore, levels of Methane (CH_4) have more than doubled since the industrial era. The studies have found evidence of a statistically significant correlation between concentrations of CO_2 in the Earth's atmosphere and temperature change [8]. Since the time of Tyndall, scientists have known that carbon-based vapours may absorb radiant heat, and this suggests causality [9]. Harvey explains how, when considering carbon-based vapours potential to absorb radiant heat it is important to consider both the degree of radiant forcing and the duration gases will remain in the atmosphere. In doing so, one arrives at estimates of what is considered a carbon-based vapours *global warming potential* [10].

Global average near-surface temperatures have warmed by about 0.7 degrees Kelvin since 1900. Over the past 30 years, global temperatures have risen rapidly at a rate of approximately 0.2 degrees Kelvin per decade reaching a level not experienced for about 12,000 years [11]. The NASA Goddard Institute for Space Studies found in 2011 that global average surface temperature continues a trend in which nine of the 10 warmest years in the modern meteorological record at that time had occurred since the year 2000. The study reports that the average temperature around the globe in 2011 was 0.51 degrees Kelvin warmer than their mid-20th century baseline [12]. The Stern Review forecasts that further rises in global average near-surface temperatures could threaten the well-being of people across the world, predicting: the melting of glaciers; declining crop yields; ocean acidification; rising sea levels; risks of malnutrition and heat stress; permanent displacement of communities; and mass extinction of between 15-40 percent of Earth's species [13].

This means that activities undertaken privately, release carbon-based vapours that contribute to a cost born by all society globally. A society as diverse in its tastes and interests, as political organisation. Faced with fear for our future, even those of just intentions may be condemned to permanent hostility whilst Public assurance is not forthcoming and apprehensions grow of the faithfulness of others. Under such circumstances, Rawls asserts that as a free and rational person one has a natural duty to comply with just institutions where they apply and assist in the establishment of just arrangements where they do not exist, at least when this might be achieved with little cost to ourselves. Along with such action comes a duty to show others mutual respect and a willingness to observe situations from their point-of-view, considering their perspective of the Good. Further, one should be prepared to give reasons for their judgments wherever the interests of others are materially affected. In this way any proposition can be offered in Good faith, enabling others to accept constraints on conduct [14].

How could one ensure their demands are of maximum social benefit?

Coase suggests that one might need to add the costs of their greenhouse gas emissions born by all society to their own private costs when taking action. This would correct production processes so they are of maximum social benefit [15]. Precedents use the term *social cost of carbon* to define the mean average costs of greenhouse gas emissions born by all society for this purpose. However, in this context it seems there are different perspectives that may be taken towards what this social benefit may be. An explanation of two prominent perspectives follow, the differences between them yielding significantly different outcomes for action.

Notable Precedents: Utilitarian Perspective

Calculation of a Social Cost of Carbon has proven to be highly rewarding for those taking utilitarian perspectives, prescribing happiness as a whole as the ultimate end to action [16]. The resources consumed in satisfying human needs require economic considerations to efficiently manage personal well-being. Normative economists often regard only individual circumstance, or level of welfare, in a state of affairs as important. This is known as the neutrality assumption [40]. Arrow, May and Sen assert that this position of neutrality is effectively an avocation of anonymity with respect to social states, which is a requirement that human beings be treated equally [17] [18] [19]. Waldron defines the notion of social welfare, an aggregate of individual welfare, as being goal-based [20]. Dworkin states that a goal can be considered a non-individuated political aim [21].

Early contributions to address this research question were awarded the Nobel Prize [22] and Life peerages in the House of Lords [23]. These notable precedents prescribe aggregate or social welfare as a goal for social benefit. Aggregate welfare is not evenly distributed amongst the worlds population. In 2019, Business Insider magazine reported that the 26 richest men had more combined wealth than the poorest 3.8 billion people on Earth [24]. Therefore, pure reinforcement of aggregate welfare is largely in the interest of a few wealthy people. This approach typically employs Frank Ramsey's Mathematical Theory of Saving as a decision-making tool. The process involves creating complex integrated assessment models of humanities social and environmental systems that apply gross assumptions of Earth's development. Mechanisms that determine the state of the world are identified and the consequences of alternative policies charted so that consequences can be valued. Welfare surpluses are then estimated for policy options, by making projections of differences from the *status quo* 100's of years into the future. Valuation of these surpluses involves setting a global social discount rate and this requires a set of personal assumptions [40]. A summary of the underlying decision-making criteria is shown follows [25]:

$$V_t = \sum_{t}^{\infty} \beta^{(\tau-t)} \cdot U(C_\tau) \quad \text{for} \quad t \geq 0 \quad (3.1)$$

Where,

$$\beta = \frac{1}{(1+\mu)} \quad (3.2)$$

Where, V_t is a *generations welfare*, β is the *discount factor*, U is *welfare*, C_τ is *consumption during time-step*, t is *time* and μ the *social discount-rate*.

$$\mu = \sigma \cdot g + \delta \quad (3.3)$$

Where, σ is the *marginal utility of consumption* and the difference in the utility one would gain from a unit of consumption by those of low and high incomes, g is the *long-term growth rate* in consumption, δ is the *pure rate of time preference* and our impatience to consume in fear of extinction.

$$S = \frac{\mu - \delta}{\sigma \cdot \mu} \quad (3.4)$$

CHAPTER 3. ETHICS

Where, S is the *savings rate* and the proportion of output that should be invested.

In practical applications, $U(C_\tau)$ can be substituted for net cash-flow in time period to yield a familiar equation.

This complex process results in considerable disagreement amongst Social Cost of Carbon estimates for a historic time period using this method. The assumptions for a global social discount rate used for three notable contributions are provided in Table 3.1. The global social discount rate for these three contributions (μ) is similar. However Cline is outlying in estimation of the difference in utility the wealthy gain from a unit of consumption compared to the poor (σ). None of the three contributions agree on a long-term global growth rate (g). When considering impatience to consume in fear of extinction (δ) it is Nordhaus that appears outlying. This yields very different values for the amount of surplus to be saved relative to that consumed amongst the three contributions (S). Clearly the worlds envisioned by just these three notable precedents are really quite different in nature and perhaps the reality is that the ethical positions of these contributors are in competition. Nobody has the power in reality to set a social discount rate in this way for all society. For most, the setting of these assumptions is of private preference only.

CHAPTER 3. ETHICS

Table 3.1: Notable Precedents: Assumptions in Social Discounting

Contribution	μ	σ	g	δ	S
Cline, 1992 [26]	0.05	1.5	0.033	0	0.67
Nordhaus, 1994 [27]	0.05	1	0.02	0.03	0.4
Stern, 2006 [13]	0.05	1	0.049	0.001	0.98

Figure 3.1 is a graphical representation of Social Cost of Carbon estimates by the Interagency Working Group on Social Cost of Carbon, United States Government. This meta-analysis considers the results of three pioneer integrated assessment models developed in the 1990's. These are the DICE, PAGE and FUND models [27] [28] [29]. The modelling simulations considered by this study returned a Social Cost of Carbon anywhere between <2007$0tC^{-1} to a 95th percentile of 2007$128tC^{-1} with μ set at 0.03 [30]. Based upon these results for a Social Cost of Carbon, it is clear that utilitarians would find it difficult to significantly alter their actions and agree on options to be taken.

CHAPTER 3. ETHICS

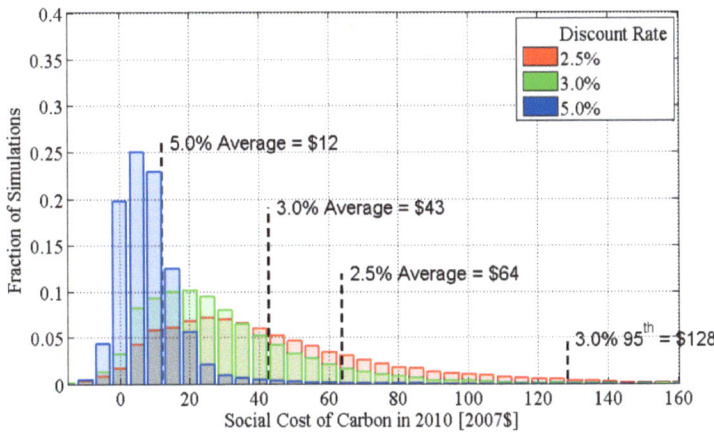

Figure 3.1: Social Cost of Carbon in 2010 (in 2007 dollars per metric ton)

> "Here there are a number of possibilities. A collective decision may determine the rate of saving while the direction of investment is left largely to individual firms competing for funds. In both a Private Property as well as in a socialist society great concern may be expressed for preventing irreversible damages and for husbanding natural resources and preserving the environment. But again either one may do rather badly." [14]

Even among welfare economists, some reject applying pure time preference so that one allows the living to take advantage of their position in time to favour their own interests [16] [25]. Such examples ignore the possibility for the living to wrong their predecessors or descendants.

CHAPTER 3. ETHICS

Dasgupta argues that it is one thing to urge that an imperfect economy should be improved, quite another to pretend that the imperfect economies we inhabit are Utopia in the way these contributions suggest. Collective saving for the future has many aspects of a Public Good, though under conditions of arising problems of isolation and assurance [31] [32]. Those who follow the principles of the social contract when taking action would probably find these methods have little relevance to their sense of justice. These contributions also do not consider the possibility for states of affairs to deliver imperfect procedural justice leading to noncompliance. Libertarian's might argue that such collective action would lead to violations of individual rights to even a slight extent and therefore should be rejected [4]. Advocates of the free-market may criticise such methodologies that involve seeking prosperity through centralised long-term coercive planning, rather than appropriate regulation that allows local agents to adjust their activities according to the present situation [33]. There is indeed nothing sacrosanct about the Public decision concerning the level of savings and its bias with respect to time preference deserves no special respect [14]. Therefore, an alternative proposition is sought.

Proposition: Hobbesian Perspective

An alternative perspective is to use the principles of the social contract to estimate a Commonwealth Cost of Carbon for decision-making. Here, one cannot rely on the assumption that the ideals of liberal democracy or social welfare are necessarily shared by all those affected. However, one might assume universal acknowledgement of a Hobbesian ecosystem.

The problem is to use Hobbesian thought to value the social cost of greenhouse gas emissions as a Good. Dasgupta asserts that this valuation need involve comparison of worlds with and without greenhouse gas emissions and stable currency [40]. Our understanding of the carbon cycle helps us determine that a world without greenhouse gas emissions is a world without Life. The Moon is an example of a world without a significant carbon cycle. Carbon appears an essential Good Life cannot do without. Our next task is to understand what a world would be like without stable currency. Hobbes would suggest this be a war of *all against all*, a world without submission to a Sovereign power. To maintain confidence, citizens of a well-ordered society will normally want the rule of law maintained. Although we might acknowledge that a common sense of justice is shared and that each wants to adhere to its arrangements, one might nevertheless lack full confidence in one another. One might suspect that others are not doing their part, and so may be tempted not to do theirs. The general awareness of these temptations may cause social systems to break down. The role of an authorised Public interpretation of rules supported by collective sanctions is needed precisely to overcome this instability. For this reason alone, a coercive State is always necessary, even though in a well-ordered society sanctions may be slight and may never need be imposed. Therefore, the penal machinery of State becomes ones security to another and is entirely sacrificial [14].

Therefore, in order to optimise ones actions to assure *faith in promise*, one arrives at a quality-of-Life indicator equivalent to a *commonwealth cost of carbon* for a given time-period, we complete the following calculation:

$$\frac{anthropogenic\ expenditure\ on\ enforcement}{anthropogenic\ greenhouse\ emissions} \quad (3.5)$$

It is evident that this calculation method requires relatively straightforward math, but the accounting is not trivial. An initial anchor for the year 2016 could be arrived at through dividing Public EU accounts for global military expenditure (\$1.8tn) and dividing by a Kyoto Protocol figure for global gross greenhouse gas emissions (10GtC) to arrive at \$180tC^{-1} [34] [35]. The genuine figure for a Commonwealth Cost of Carbon might be double this initial anchor, once global expenditure on Public order and safety were included in the gross enforcement cost total, ranging somewhere between \$300tC^{-1} and \$400tC^{-1} [36]. Unfortunately, accurate figures for these expenditures remain difficult to obtain and involves considerable uncertainty. This leads to the belief that estimation of a Commonwealth Cost of Carbon using the principles of the social contract return results of a different order of magnitude to even the higher utilitarian estimates in 2010 that average 2007\$64tC^{-1} with μ set at 0.025 [30].

There are a number of implications in taking a Hobbesian approach to calculating the Commonwealth Cost of Carbon which are very different from the utilitarian perspective. Here, a Commonwealth Cost of Carbon becomes of no concern when there is no need for anyone to enforce a promise - a Hobbesian Utopia that is very difficult to achieve. Further, maintaining enforcement costs, whilst

CHAPTER 3. ETHICS

reducing greenhouse gas emissions, increases the Commonweath Cost of Carbon for those following the principles of the social contract for the global commons. The correction to decision-making required appears substantial and may be applied to every personal cost-based decision of anybody. There is no reliance on an ideal observer, all peoples are treated equally and allocations of property rights have no influence on the ability of anyone to estimate. The Hobbesian perspective does not support the view that anybody should be embargoed for their contributions to climate change. It also does not indicate a target global mean surface temperature rise, as this is no longer the chief concern.

It is assumed that each person reasonably believes that others have a sense of justice and an effective desire to carry out their obligations. Without this mutual confidence, little is accomplished by uttering words. The indivisibility and Public nature of certain essential Goods, such as the element Carbon and the State of the Earth, give rise to externalities and temptations that necessitate collective agreements organised and enforced by states. Once Goods are indivisible over large numbers of individuals, their actions decided upon in isolation from one another will not lead to the general Good. Some collective arrangement is necessary and everyone wants assurance that it will be adhered to if willing to take part. In a large community, confidence in one another's integrity rendering enforcement superfluous is not to be expected [14]. Recognition of a Commonwealth Cost of Carbon involves dealing with a system of transaction costs and might require a highly centralised intervention. Ideally, this Commonwealth Cost of Carbon might be applied as a common rate of value-added tax on the global warming potential of delivered fuels, including food, to be excised as duty by a society. Alternatively, the entire society could be internalised within a single non-coercive firm [15]. Proportional expenditure taxes have been noted as potentially being part of a best tax scheme that treats everyone in a uniform way, with the possibility of making exemptions for dependents [37]. Either option could be open using the Hobbesian approach, as the relatively simple calculation methodology involved may be an auditable record by the general Public to overcome major assurance problems effecting the stability of any covenant. A remaining challenge exists in assuring that all parties have the necessary assurance to follow through on such commitments. This involves members giving promise and a reciprocal recognition of their intention to put themselves under an obligation that is to be honoured. Through reciprocal recognition and common knowledge such arrangements can be enabled, started and preserved [38].

By choosing a higher and Publicly auditable Commonwealth Cost of Carbon a broader range of interventions that lead to greenhouse gas abatement become viable, which may include: closer activity control to reduce waste; reduced development of excessive plant and infrastructure capacity; increased utilisation of productive assets; renewable energy sources; and active modes of transport, such as walking and cycling. The overall effect would be to promote a society that makes solemn recognition of the great sacrifices made to maintain the current global State of affairs, observing more diligent attention to their own impacts upon the commons.

3.4 Minimum Decency

Regardless of ones goals and concerns for either the welfare of an individual or society as a whole, there is a requirement for all humanity to be treated with minimum dignity and respect. Such perspectives of human dignity and respect need to relate to a common sense of justice and fair priority shared fairly by all. Rawls' [14] full statement of justice for institutions is a concise and widely recognised solution to the priority problem of justice as fairness. A sense of which could be shared and applied. An account, but not a defense, follows.

First Principle

> "Each person is to have an equal right to the most extensive total system of equal basic liberties compatible with a similar system of liberty for all."

Second Principle

> "Social and economic inequalities are to be arranged so that they are both:
>
> a) to the greatest benefit of the least advantaged, consistent with the just savings principle, and
>
> b) attached to offices and positions open to all under conditions of fair equality of opportunity."

First Priority Rule: The Priority of Liberty

"The principles of justice are to be ranked in lexical order and therefore liberty can be restricted only for the sake of liberty. There are two cases:

- **a)** a less extensive liberty must strengthen the total system of liberty shared by all;
- **b)** a less than equal liberty must be acceptable to those with lesser liberty."

Second Priority Rule: The Priority of Justice over Efficiency and Welfare

"The second principle of justice is lexically prior to the principle of efficiency and to that of maximising the sum of advantages; and fair opportunity is prior to the difference principle. There are two cases:

- **a)** an inequality of opportunity must enhance the opportunities of those with the lesser opportunity;
- **b)** an excessive rate of saving must on balance mitigate the burden of those bearing the hardship."

General Conception

"All social primary goods—liberty and opportunity, income and wealth, and the bases of self-respect—are to be distributed equally unless an unequal distribution of any or all of these goods is to the advantage of the least favoured."

3.5 Rate of Saving

Rates of saving typically relate to the percentage of disposable income saved, rather than spent, during a time-period. Equally the term also relates to the percentage of financial return an investment. By its very nature, considerations towards rates of saving have both implications for the balance of equity between generations and associations. As an aggregation of the well-being of individuals, calculations of rates of saving are often described as *inter-generational well-being*. Rawls' second priority rule stipulates a requirement to control rates of saving, so that excessive rates on balance mitigate the burden of those bearing the hardship. Therefore, there is a requirement to estimate the present generations rate of saving.

Cost-Benefit Analysis

Dasgupta [40] explains how we *value* when comparing objects and we *evaluate* when comparing the benefits of actions. Valuation and evaluation both involve comparisons between worlds with and without the course of action or object. This process is called social cost-benefit analysis and involves measuring consumer and producer *surpluses*. Carrying out social cost-benefit analysis requires a quantitative formulation of inter-generational well-being, for which Ramsey's Mathematical Theory of Saving [25] is a well respected candidate.

Accounting

There are a broad range of influences on well-being that need to be accounted for. An original taxonomy is illustrated in Figure 3.2, inspired somewhat by Dasgupta [41].

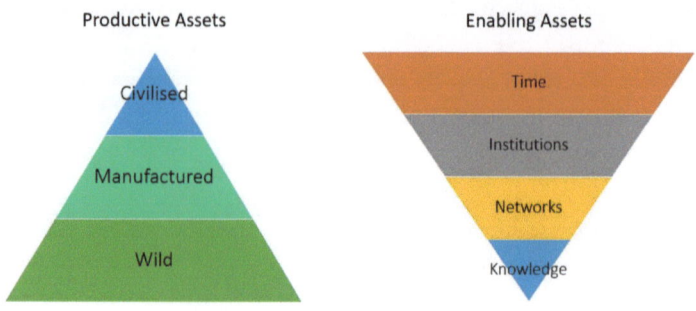

Figure 3.2: Original taxonomy of influences on human well-being

Enabling assets allow society to be more precise about their influences on the physical environment, may be defined as software and possibly scale. Dasgupta [40] describes how enabling and production assets may offer different value:

Use-value: the contribution of an asset to well-being once consumed;

Intrinsic-value: the contribution of an asset to well-being in the understanding that it exists (e.g. believing that polar bears roam free in the arctic);

CHAPTER 3. ETHICS

Option-value: the contribution of an asset to well-being in the understanding that it may be consumed in the future.

The attitudes and preferences that characterise a state of well-being are not defined and need to be found. Recognised methods to evaluate social attitudes and preferences are:

Stock markets: Determines the values of a supply of commodities through a process of market clearing.

Hedonic regression: Determines the value of amenities by comparing house prices and the amenities available on different parcels of land to each other using regression analysis.

Satisfaction/approval surveys: Determines level of approval from a representative sample of respondents for a particular amenity. Arrow et al. [39] recognised that respondent heuristics prevent valuation of an amenity this way.

Limitations

Wild capital, by its very nature, does not submit to any social contract governing states (a pre-requisite for currency). Therefore, no human can confidently place a valuation on Wild Capital and it's contributions to well-being may be priceless. Cost-benefit analysis involves a number of personal assumptions that may not be particularly relevant to many people. This tool for decision-making is best suited to personal savings behaviour. Cost-benefit analysis may be entirely inappropriate for decisions relating to *essential Goods*. Attention to distribution outliers and qualitative information is essential in assuring a just approach to personal savings goals.

3.6 A Note on Perfectionism

There is a body of thought that define the duty and obligations of individuals so as to maximise the achievement of human excellence in art, science and culture. Neitzsche [42] asserts:

> "Mankind must work continually to produce individual great human beings—this and nothing else is the task... for the question is this: how can your life, the individual life, retain the highest value, the deepest significance?... Only by your living for the good of the rarest and most valuable specimens."

However, in order to arrive at perfectionism, Rawls [14] asserts that we would have to attribute all parties to a prior acceptance of a certain style and aesthetic grace, and to advance the pursuit of knowledge and the cultivation of the arts. Whilst values of excellence may be recognised, human principles need to be pursued within the limits of the principle of free association. The coercive apparatus of states would not be used to win greater liberty or larger shares of wealth on the grounds that one's activities are of more intrinsic value. The social resources necessary to support associations dedicated to advancing the arts, sciences and culture generally are to be won as a fair return for services rendered, or from such voluntary contributions as citizens wish to make. Governments might limit support to cases of overcoming isolation and assurance.

3.7 Demonstration, Assurance and Automation

It is believed that the arrangements explained pose fair arguments for an equal liberty of conscience in managing the global commons. Therefore, parties have Good grounds for adoption. These arguments allow the choice of a regime of moral liberty and freedom of thought and belief, and of religious practice, although these may be regulated as always by state interest in Public order and security. Particular associations may be freely organised as their members wish, and could have their own internal Life and discipline. Acceptance of these limitations does not imply that Public interests are in any sense superior to moral and religious interests, nor does it require that government hold indifference to religious matters or hold the right to suppress philosophical beliefs when in conflict with state affairs. Associations represent demonstrations of different interest and may be genuine causes of further alarm in need of address. By exercising powers in this way governments may act as the citizens' agent and satisfy their demands of a Public conception of justice. Maintenance of Public order is understood as a necessary condition for everyone's achieving ends whatever they are and to fulfil their interpretation of moral and religious obligations. A reliance on particular metaphysical doctrine or theory of knowledge is not required. The appeal is made to common sense, plain facts

CHAPTER 3. ETHICS 34

and reasoning available to all. When the denial of liberty is justified by an appeal to Public order as evidenced by common sense, it is always possible to urge that the limits have been drawn incorrectly, and that experience does not justify the restriction.

Capital budgeting involves a keen understanding of demand, so that plans can be made that develop affordable and secure supply-chains with an appropriate degree of quality assurance. One needs to strike a balance between quality and the activities promoted through what is demanded by considering the Commonwealth Cost of Carbon. Professional best practice would be to add the Commonwealth Cost of Carbon to expected fuel costs to budget for each items assurance requirements. *Uncaught promises* become a high priority focus area in all disciplines of engineering to improve integrated systems stability. Any policies that stipulate objectively under-assured supply-chains runs risks of severe problems and challenge. The consequences of failing to perform to shared understanding are simply increased levels of enforcement by states, whether judicial, or ultimately through military intervention.

CHAPTER 4

Continuous Integration

To support a diverse market of complementors platform ecosystems define boundaries for developers to integrate and offer new Features to consumers. The Guidance and Standards that constitute these boundaries provide criteria for failure to any developer and sets a feedback-loop for learning. Whilst *Guidance* criteria are suggested as Good practice for learning practitioners, *Standards* represent mandatory criteria where it appears no other reasonable option is available. This process of development and learning to meet integration criteria is known as a process of continuous integration. In the interest of the least-advantaged developer, continuous integration processes need offer rapid results with instructive and relevant feedback. Developers need the ability to frequently test Features for compliance with integration requirements and indication of whether they have achieved desired outcomes.

Guidance, Standards and review processes relating to the compliance criteria of Features may be stipulated at various stages of Feature development. Features that exhibit a development stage need test against the relevant criteria. These development stages are then mapped across system classes with significantly different requirements to establish a matrix for designating test criteria. System classes are not mutually exclusive and any Feature may exhibit none, one or many of them. Test development is an important consideration of any *Framework*, which sets system definitions and dependencies. It is important that developers recognise the system classes which are relevant to any Features. An outline matrix for designating test criteria is shown in Table 4.1.

Table 4.1: Outline matrix designating test criteria

Class/Stage	life	alarm	property	sensitive	personal	client	public
use							
architecture							
framework							
device							
system							
integration							
acceptance							
penetration							
market							

What follows is a brief summary of some key criteria that has been established for each system class. The intention is that these criteria represent a pragmatic minimum for ecosystem development. There is a reluctance to be over-prescriptive in order to maintain as extensive a system of liberty as possible.

4.1 Life

A *life* system maintains conditions of minimum decency. The risk of Life systems failures may have direct impacts upon health. It may be perceived Professionally negligent to develop conditions that knowingly increase risk of failure in a Life system without making best efforts of mitigation.

This investigation has identified a need to comply with a Hobbesian Framework for Life [5]. Such a Hobbesian perspective leads towards a Commonwealth Cost of Carbon as an appropriate quality of Life standard for this platform ecosystem at the market stage. The Earth itself, as a candidate Sovereign of State over the commons, has ultimate responsibility to apply enforcement however one wills. It is understood that no Sovereign of state, in the sense of any contracted protective association, has the power to enforce its rules over all Life on Earth and no platform can exist without promises assured by at least one state.

Life systems tend to be most promising when relying on simple, clear common-sense processes with little in the way of conflicting interests. A defining feature of Life systems is that integration with alarm managers is restricted to hard-wired digital inputs only. This is to assure no interference from less critical systems over more permissive networks.

4.2 Public

There are many reasons why it may be beneficial for an alarm management system to be exposed to *public* networks. These include:

- inter-operability with web services;
- connection to remote workstations;
- monitoring by system manufacturers;
- monitoring by remote plant historian.

System level protocols may be quite permissive to support an extensive range of options for development. Therefore, *private* systems may allow for a diverse and risky mix of traffic that include unencrypted broadcasts which have not sufficient security features for inter-networking. Network Address Translation (NAT) is a service that operates on a router to connect Private networks to Public networks for inter-networking. NAT is often implemented at the Wide Area Network (WAN) edge router to enable internet access across sites. With NAT, an organisation needs one Internet Protocol (IP) address to represent an entire group of devices as they connect outside their domain. Port Address Translation (PAT) enables one single IP address to be shared by multiple hosts through port forwarding of both IP and port addresses using IP packet filter rules. NAT is a networking feature that can help reduce organisational security risk by hiding internal networks from Public networks. By default, outside Public IPs cannot communicate to internal Private IP devices if there is no pre-existing NAT translation. Connections between internal Private devices and outside Public devices must be initiated by outbound communication. So, NAT separates Public and Private networks. Additionally, organisations that use NAT can implement and maintain multilayer security to block threats and protect malicious activity.

A typical network configuration for an alarm management system bound to on-premise communications is illustrated in Figure 4.1.

Following Guidance for Public network criteria, an acceptable proposition for integration of this Private network with remote clients is illustrated in Figure 4.2.

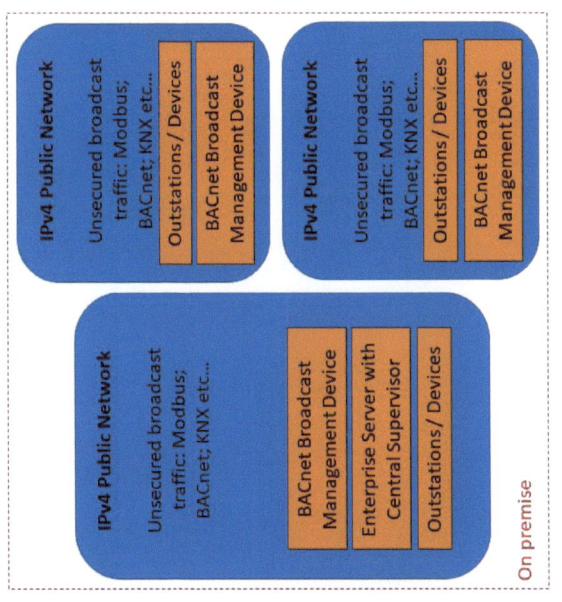

Figure 4.1: Typical network configuration for on premise local alarm management system

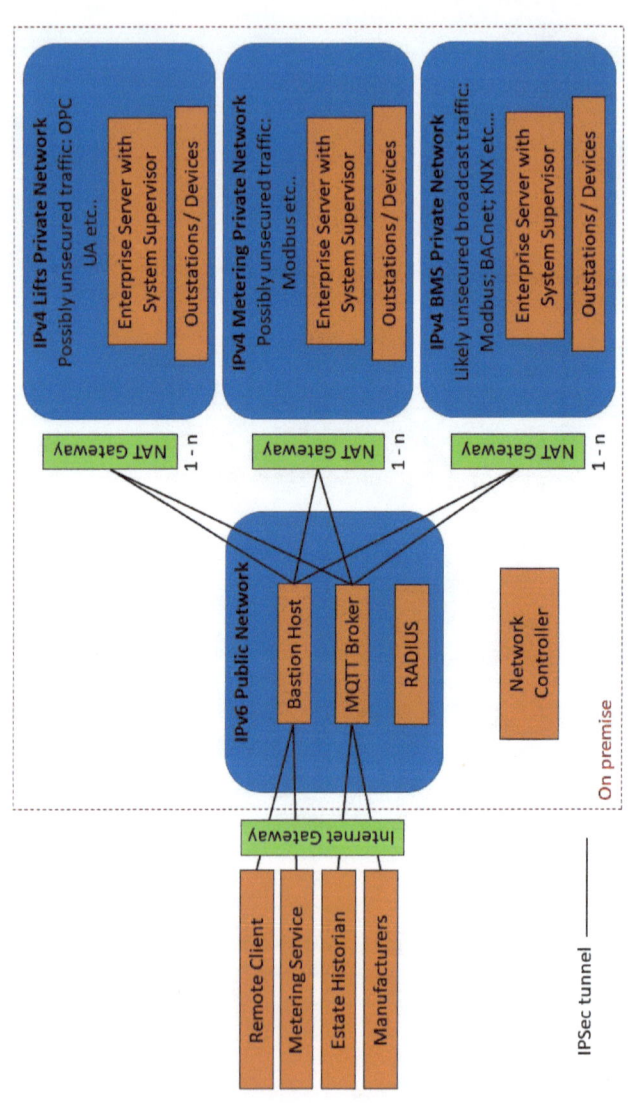

Figure 4.2: Acceptable proposition for integration of Private alarm management system with wide area network

CHAPTER 4. CONTINUOUS INTEGRATION 43

Demilitarised Zone

IPv6 is the most recent version of IP designed to supply IP addressing and additional security to support expected growth in connected devices in IoT, manufacturing and emerging areas like autonomous driving. The primary reason to implement IPv6 is addressing. IPv4 is based on 32-bit addressing, hence, limited to 4.3billion addresses. IPv6 is based on 128-bit addressing and can support 340trillion3 addresses. IPv6 provides more than enough globally unique IP addresses to network every device on the planet, helping ensure networks can keep pace with the expected proliferation of IP-based devices [43].

A feature of IPv6 is that the protocol optionally implements IPsec security in either transport or tunnel mode. IPsec in transport mode encrypts the transport layer of the IPv6 datagram, whilst IPsec in tunnel mode encrypts the entire IPv6 datagram. Therefore, use of IPv6 along with IPsec tunnel mode allows for a more secure implementation for Public Demilitarised Zones (DMZs).

While IPv6 offers a large number of static IP address space to fulfil increasing host demands, chances are systems that include field layer devices will require IPv4 and IPv6 addresses to coexist. NAT allows organisations to connect IPv6 and IPv4 networks using NAT64 translations. Such translations may be:

- 1-n, where a NAT Gateway exposes a single IPv6 address to the Public and each IPv4 address within the Private network is represented by a designated port;

- 1-1, where a NAT Gateway exposes one IPv6 address

to the Public for each IPv4 address within the Private network.

Remote Authentication Dial-In User Service (RADIUS) is an authentication protocol that grants or denies users access to a network's services. When a user tries to access a protected source, the RADIUS confirms they should have access first, by checking their credentials or certificate, before rejecting or authenticating users accordingly. RADIUS requires the following components:

- a RADIUS server;
- a directory of user/device information;
- a RADIUS client.

RADIUS employs user credentials or certificates for role-based access control, which involves the checking that a user is authorized and granting an appropriate level of privilege to the network. The use of digital certificates for IEEE 802.1X role-based authentication by RADIUS is considered more secure than user credentials (username and password) [44]. It is thought RADIUS is most successful when authenticating users with IEEE X.509 type digital certificates. [45]

A bastion host is a server whose purpose is to provide access to a Private network from an external network, such as the internet. Because of its exposure to potential attack, bastion hosts are designed to minimise chances of penetration. Bastion hosts are a critical point of network security, hence, it is best practice to harden such instances. Examples of hardening might be to disable unnecessary applications or services, tuning the network, regular monitoring

CHAPTER 4. CONTINUOUS INTEGRATION

with audit tools and multi-factor authentication of users (by RADIUS, for example).

Message Queued Telemetry Transport (MQTT) is the dominant standard used in Internet of Things (IoT) communications. Because of its simplicity, MQTT doesn't require much processing or battery power from devices. With the ability to use very small message headers, MQTT doesn't demand much bandwidth either. MQTT also make it possible to define different quality of service levels for messages that enable controls over how many times messages are sent and what handshakes are required to complete them. The core of the MQTT protocol are clients and servers that send many-to-many communications between multiple clients using the following:

- Topics, which provide a way of categorising the messages sent;

- Publishers, include the devices configured to send out messages containing data with associated topics;

- Subscribers, include the devices that receive data related to pre-defined topics;

- Brokers, which receive published messages and distribute to clients subscribed to the specific topic. These brokers may store a work queue awaiting distribution to subscribed clients.

Message brokers may be deployed on virtual machines or containerised (Docker/Kubernetes) environments [46].

An internet gateway allows communication between devices and the internet, if provided with Public IPv4 or IPv6 addresses. Similarly, devices on the internet can initiate a connection to resources with Public IPv4 and IPv6 addresses over an internet gateway. An internet gateway may be used in conjunction with a firewall to restrict inbound and/or outbound connections from specific Public IP address ranges.

Network Control

Software Defined Networks (SDN) is a service that makes it possible to manage a network via software. SDN enables separation of a control plane for setting data processing rules and a data plane responsible for processing data carrying packets accordingly. This system architecture allows for automation and programmability of the network. It is a centralised system that permits management from a single network controller, for which the controllers security is absolutely critical to the performance of the entire local area network. There are various SDN protocols available.

OpenFlow allows switches from different vendors with proprietary interfaces to be managed remotely over a single, open protocol. The protocol allows remote and programmable administration of a layer 3 switches packet forwarding tables by adding, modifying and removing matching rules and actions. This allows routing decisions to be made periodically or at a necessary moment by the controller, which are then deployed to a switch's flow table. The actual forwarding of matched packets is then left to the switch. The OpenFlow protocol is layered on top of the Transmission Control Protocol (TCP), with controllers listening on port 6653 for switches Address Resolution Protocol (ARP) requests.

CHAPTER 4. CONTINUOUS INTEGRATION 47

Open vSwitch is an open source implementation of a distributed virtual multilayer swtich. Its main purpose is to provide a switching stack for a hardware virtualisation environment, whilst supporting multiple protocols and standards. The protocol is designed to support transparent distribution across multiple physical servers by enabling creation of cross-server switches in a way that abstracts out the underlying server architecture. The projects source code is distributed under the Apache-2.0 license.

Automated Testing

The use of SDN network controllers develops an opportunity to employ device, system and integrated system testing via automated tests administered over the network infrastructure. In doing so, the entire installation can be tested for compliance against specific criteria, from device on-boarding, through to full operation.

Device Automated Qualification (DAQ) for IoT devices is a framework designed to test and operate IoT devices within an enterprise environment. Concerned with device testing and qualification, DAQ provides a means to automate many device enrollment testing requirements. This results in a more manageable, robust and secure platform. Distributed under the Apache-2.0 license, DAQ is based on the FAUCET OpenFlow network controller and has the following capabilities [49]:

- Automated device qualification and testing, through standardised tests of device behaviour against established security and network standards;

- Use of the FAUCET OpenFlow controller to orchestrate *micro-segmentation* of a network to improve security.

Nmap is an open source utility for network discovery and auditing. Many systems and network administrators also find it useful for tasks such as network inventory, managing service upgrade schedules, and monitoring host or service up-time. Nmap uses raw IP packets in novel ways to determine what hosts are available on the network, what services those hosts are using, what operating system they are running, what type of filters/firewalls are in use and dozens of other characteristics. Designed to rapidly scan large networks, Nmap may also be used to scan a single host. Nmap is distributed on GNU GPLv2 license and runs on all major computer operating systems [50].

Ncrack is a high-speed network authentication cracking tool. It has been built to help companies secure their networks by proactively testing all their hosts and networking devices for poor passwords. Ncrack's features include a very flexible interface granting the user full control of network operations, allowing for very sophisticated bruteforcing attacks, timing templates for ease of use and runtime interaction similar to Nmap. The software is distributed on GNU GPLv2 license and runs on all major computer operating systems. Ncrack supports many of the most widely used protocols, including [51]:

- Secure Shell (SSH);
- Remote Desktop Protocol (RDP);
- File Transfer Protocol (FTP);
- Telnet;

- Hypertext Transfer Protocol (HTTP);
- MQTT;
- MySQL;
- MongoDB;

4.3 Property

A *Property* constitutes an assured promise. Assurance of Property is enforced by states and is subject to a state's particular legal Framework. Unless some sort of agreement has been made between states, there is no guarantee that assured promises may be understood similarly across legal Frameworks. It is understood that the promise of intellectual Property applied to this document is assured by a Framework defined in English Law. The present Standard for the diligent management of Property are version control systems, for which Git is a popular case [52]. With any promise of Property at any development stage, it is essential for assurance purposes to define contractual terms for its proper use under a License.

4.4 Alarm

An alarm management system constitutes processes that effectively discharge demands on a governance regime. Alarm managers may take into account field observations and define procedures in order to take appropriate actions. They may be situated in outstations within the field and procedures defined by scripts. This could result in alarm managers being constituted as operating computational controllers. More resilient and maintainable outcomes can be achieved by keeping supervisory control of a specific alarm management system of outstations within a single Private network

Having adopted a Hobbesian perspective of the ecosystem, a Framework for alarm management defined by John Rawls, A Theory of Justice has been followed [14]. Alarm management functions need to be Professionally reviewed by domain experts following this platforms Rawlsian Executive Function as a reasonable minimum. Such alarm management reviews require understanding of market demands and demonstrations of different interests through the development of associations. Rulings that appropriately define alarm manager operations, in the interest of timely and consistent responses, would become a promise of Property. Promising rulings would require a system of Professional peer-reviews, with any proposed solution represented by a formal domain Ontology and compliance with the principles of this Executive Function as a minimum. Continuous integration products, such as Jenkins or Travis CI, may be used for automated testing of software and rapid development feedback in preparation for operational releases [55] [56]. It is important to allow for full point-to-point testing of an alarm management system during an in-situ commissioning phase following full deployment.

Alarm managers in the field may use a diverse range of network and hard-wired interfaces. Alarms need be priori-

CHAPTER 4. CONTINUOUS INTEGRATION 51

tised in a common-sense manner to direct attention:

- P1, immediate action;
- P2, urgent action;
- P3, pressing action;
- P4, optimisation action.

4.5 Sensitive

Sensitive information may be subject to assured promises that require individuals or associations to retain information in Private. Such information would likely be subject to a Framework defined by a state. Sensitive information should be encrypted with AES data encryption Standards at rest [53]. Encryption in transit can be achieved over Private IPv4 networks through implementing Transport Layer Security Protocol (TLS) [54]. On Public networks, encryption in transit may be provided by IPv6 tunnels.

4.6 Personal

Any information that may be used by organisations to identify individuals has particular relevance to personal privacy and security. The collection, storage and use of personal information that may be used to identify individuals is often subject to a particular legal Framework enforced by a state. Unless some sort of agreement has been made between states, there is no confidence of common requirements across legal Frameworks. Any personal information contained within this document is subject to a Framework defined in English Law. Personal information that is to be assured for Private use only should be encrypted with AES data encryption Standards at rest [53]. Encryption in transit can be achieved over Private IPv4 networks through implementing Transport Layer Security Protocol (TLS) [54]. On Public networks, encryption in transit may be provided by IPv6 tunnels.

4.7 Client

Client applications involve the release of consumer interfaces to remote devices. Development of releases require the adoption of a desirable Framework, such as Node.js, Flutter or Flask [57] [58] [59]. The sources of software released to client devices may be developed by agencies and have a promise of Property. Continuous integration products, such as Jenkins or Travis CI, may be used for automated testing of source code and rapid development feedback in preparation for operational releases [55] [56]. Client devices are typically networked over internet or serial protocols.

CHAPTER 5

Operation

The potential market of the platform ecosystem outlined here could be global. At this time, few organisations meet the criteria that define the boundaries of the platform ecosystem defined. This means there is considerable work to be done to signal failure to potential complementors and initiate feedback to support an integration process. It is also understood that the originators of this document are unlikely to be aware of all relevant information in this domain, which may result in some further change [60]. Therefore, an operational model is necessary.

Within a particularly complex and generally disinterested political context, there appears a need to establish a stable and Public auditable signal of thought that may be shared to establish mutual understanding. Such a signal may become a reliable target for further development. The following arrangements have been made to support operation of this platform ecosystem.

5.1 Agenda

A prioritised agenda for systems engineering within the domain of civil infrastructure follows. The agenda includes a set of alarms which are believed to be reasonably focussed and have some universal relevance. Investigations of each alarm are not nearly exhaustive. The intentions of recommended actions are to apply practical solutions that could quickly mitigate the most severe or prevalent symptoms. Failure to action initiatives to mitigate these risks will contribute to greater ecosystem instability.

P1: Technical Support for Supervisory Control and Data Acquisition

Infrastructure fulfilling critical roles in society often employ Supervisory Control and Data Acquisition (SCADA) tools to configure, monitor and analyse complex systems and processes. In many cases, a lack of technical support is making such systems very difficult to maintain. SCADA is more vulnerable when employing proprietary commmunications protocols and obselete operating systems. To access the broadest and most diverse business ecosystem possible for support, operators need migrate SCADA to current industry standard open communications protocols and operating systems with Long Term Support (LTS).

Further, systems manufacturers providing monitoring

CHAPTER 5. OPERATION 55

capability need to support Good practice for Public internetworking. This can allow customers to have full visibility and assured compliance of all data acquiring devices on-site.

P1: Social Media and Influencing the Averages

There is a widespread tendency to Publicly report averages, totals and probability distributions in summary analysis of performance. Excessive use of social media to influence popular attitudes and preferences in Public re-enforces these problems. Under such regimes, skill and care needs to be taken to promote correct responses, as opposed to what might actually lead to development of a confused herd.

To resolve any Public appeal to ones satisfaction, evidence of qualified judgement is required. Most of the time there is more attractive opportunity for development in considering the outliers and the qualitative. Returning to more personal engagements can be beneficial, so that a diverse collection of backgrounds and perspectives may be considered. In doing so, constructive conversations can be maintained in Private and more considerate outcomes put into practice.

P1: Compressor Failures under Minimum Turndown

Hot and chilled water systems are a common means of distributing thermal energy to terminal devices that heat and/or cool occupied spaces. When considering larger buildings or heat networks it is popular to install equipment for evaporation and/or condensing of fluids that employ

CHAPTER 5. OPERATION

compressors for pressurisation. Heat pumps and air-cooled chillers are examples of such equipment. However, it is becoming common to witness such compressors failing within a few years due to frequent start/stop cycles. This can be caused by temperature differentials between secondary flow and return temperatures narrowing under minimum turndown conditions if systems are poorly designed.

Figure 5.1 illustrates an example poorly designed chilled water system where chiller compressors are vulnerable under minimum turndown. Here the chilled water system has a common header and constant temperature secondary loop, with circulation pumps enabled by a space cooling demand signal to operate at a constant speed. In this case, when only a small proportion of terminal unit cooling control valves open, return chilled water temperatures may remain only very slightly higher than flow temperatures. This can result in marginal loads on a single duty chiller, leading to problems of frequent cycling. Failure of the chiller compressors would result in interruption of the entire air-conditioning system.

Figure 5.2 illustrates a well designed system that has not the same vulnerabilities. This chilled water system has a split header, with a flow-meter fitted to a secondary bypass. The secondary circulation pumps are variable flow and target a set-point pressure differential across the index secondary loop run by implementing proportional and integral control to modulate pump speed. Lead and lag chillers are staged on and off so that the flow of chilled water through the bypass flow-meter is always between 10 percent and 100 percent of one lag chillers capacity. There are a number of ways this configuration improves resilience:

- Failure of one smaller unit does not necessarily interrupt air-conditioning, as some other units are available to at least meet part design load;

- Through staging a number of smaller units, minimum turndown of the primary system is lower and the lead unit can be sized specifically to comfortably meet minimum load;

- Differential pressure control of the secondary pumps better maintains differences between flow and return temperatures.

When employing heat pumps for heat rejection/generation, it is often necessary for return temperatures to be closely controlled. This can be achieved by adding buffering to return pipework.

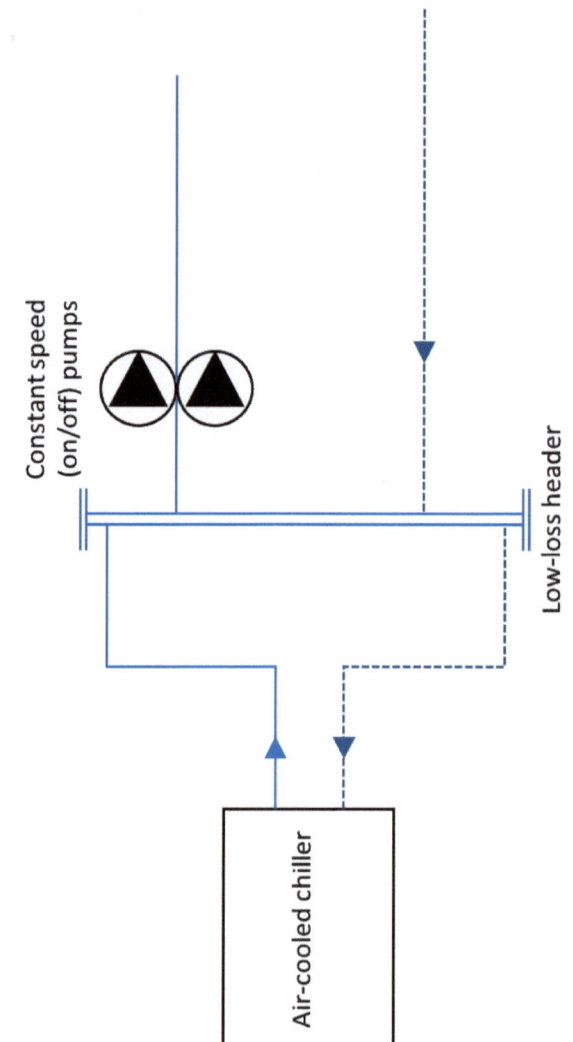

Figure 5.1: Typical common header chiller configuration

CHAPTER 5. OPERATION

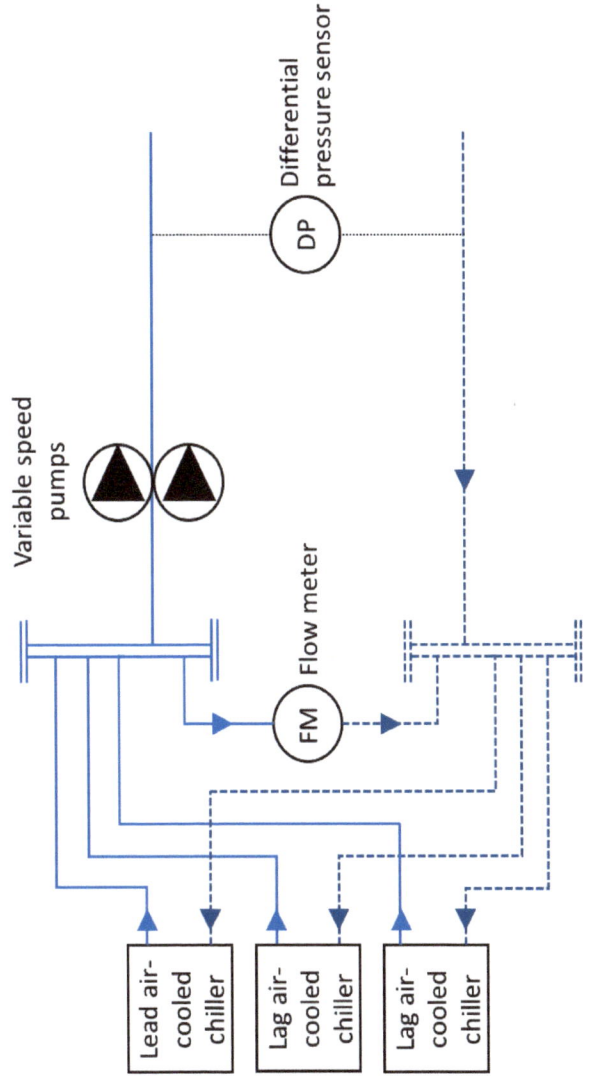

Figure 5.2: Split header chiller configuration

P2: Summertime Domestic Overheating

Evidence suggests that temperature extremes can exacerbate chronic conditions, including: cardiovascular; respiratory; cerebrovascular diseases; and diabetes-related conditions. The summer of 2022 was the joint warmest summer on record for England according to mean temperature. Four of the last five warmest summers recorded in England have occured since 2003. Between June and August 2022, there were over 3,700 excess deaths, out of a total of 56,300 deaths in England and Wales. This number of excess deaths is 6.2% above the 5-year average.

Across the world, the summer months of 2023 have also been unusually hot. By mid-July, temperatures in mainland Europe reached 46°C. The Sanbao township recorded a peak temperature of over 52°C and a peak temperature of over 54°C was recorded in Death Valley, California.

Therefore, it is becoming common for people to prioritise reducing exposure to heat stress during summer months. Air-conditioned workplaces and bathing in coastal areas may be a popular solution. However, people still tend to spend a large proportion of their time enjoying the privacy of their homes. This is particularly the case when remote or hybrid working. Therefore, there is a need to identify some practical solutions for widespread initiatives to mitigate domestic overheating.

Here, we can take some inspiration from the venacular architecture of the Mediterranean region. External shading devices are amongst the most effective tools for mitigating overheating, by blocking ingress of direct sunlight within a dwellings perimeter. Deeper window surrounds and external shutters can be particularly worthwhile. Window trellises can make for an attractive Feature.

Where security and acoustics allow, opening of windows can also be useful. The internal surfaces of thermal mass within a dwelling can be cooled if windows are left open overnight. During the day the thermal mass can then keep occupied spaces cool by absorbing extra heat as it warms up during the day. Increasing air-velocity within dwellings improve thermal comfort during summer, even where indoor operating temperatures are not significantly changed. This is because occupant thermal comfort is the combined outcome of operating temperature, humidity, insulating layers, activity levels and air-velocity. There is usually a difference in static pressure across aspects of a property. In a residence of multiple aspects, doors and windows can be opened to set up an air path between aspects. This will then achieve higher air velocities within the dwelling through cross-ventilation.

P2: Close Control of Domestic Heating

Economic activity has a strong causal relationship with energy consumption. Historically, energy prices are closely associated with overall consumer prices. Therefore, in times of inflation in consumer prices, domestic energy efficiency represents a particularly focussed investment to moderate demand. Such intervention can be beneficial to all society when applied equally to the wealthy or those in more marginal circumstances. Therefore, such an initiative may be a candidate for support by Public policy.

Radiators are widespread terminal devices used for maintaining thermal comfort by heating spaces within homes. Typically, hot water is pumped through radiator circuits at a constant volume and temperature whenever a thermostat signals space temperatures below a set-point value. Radiators are fitted with Thermostatic Radiator Valves (TRVs) to adjust the volume flow rate of hot water through the radiator when heating spaces to desired temperatures. With this configuration, if hot water flow temperature set-points or a number of TRVs are adjusted, system pressures and balance can fluctuate leading to a situation where none of the radiators within a system have TRVs set up as needed. These problems can be overcome by introducing differential pressure control to balance the heating system. This can be in the form of adding a single Differential Pressure Control Valve (DPCV) to the common return pipework of a two-pipe system. Alternatively, each radiator TRV could be replaced with a Pressure Independent Thermostatic Radiator Valve for a similar effect. Such adjustments allow radiator systems to be controlled on the basis of a set-point space temperature, which may have advantages if opting for actuated TRVs and smart digital heating adjustment via Client devices. The outcome would be a more energy efficient and comfortable roperty during winter.

P2: Availability of Delivered Electricity

There are physical limits to the quantity and stability of renewable energy sources available for distribution to consumers. Such constraints have implications for any desired widespread transition towards electric powered vehicles, buildings and infrastructure. Whilst renewable generation tends to fluctuate with the weather, quick-response Combined Cycle Gas Turbines (CCGT) and constant nuclear power remain necessary options to meet essential demands. CCGT power is particularly useful for meeting momentary differences between forecast demand for electricity and available supply, for instance during hot still summer working hours. In order for renewable energy to satisfy a larger proportion of overall end-uses, there is an urgent need to increase capacity of energy storage and manipulate energy demands.

National Grid have released delivered electricity carbon intensity factors through an Application Programming Interface (API) to support dynamic load manipulation by customers. The API provides regional forecasts of delivered electricity carbon intensity up to 24-hours in advance. Illustrations of temporal and regional variations in delivered electricity carbon intensity are shown in Figure 5.3 and Figure 5.4 [61]. With such innovation, there is an opportunity for energy centres to charge or discharge thermal stores and/or batteries according to expected times when conditions affecting electricity supply and demand are most favourable. This requires system controls to be exposed Publicly to the WAN, requiring skill and care. There is also an opportunity to procure transport and equipment according to regional characteristics of electricity distribution, rather than considering entire nation states as being similar. For example, at the time of writing there is a stronger business case for the installation of heat pumps to provide thermal comfort in Scotland than in Cornwall.

Figure 5.3: Carbon intensity forecast

CHAPTER 5. OPERATION

Figure 5.4: Regional carbon intensity

P2: Air Handling Unit Coil Failure During Frosts

Typically, larger buildings require air handling units to provide a minimum quantity of fresh air to occupied spaces to maintain indoor air quality. Such air handling units may have supply and extract ducts, each fitted with fans and heat recovery. Supply ducts are also often fitted with heating and cooling coils, filters, an array of instrumentation and dampers. Coils are susceptable to frost conditions during the winter season. Without appropriate control of equipment, it is common for such coils to crack and fail when air handling units operate in cold conditions.

To mitigate such risks, it is important for air-handling units to be fitted with frost stats and operate frost routines. Off-coil capillary tube frost stats should be laced across the initial heating coil meeting incoming fresh air, whether that be a dedicated frost coil or heating coil. Frost stats should be set to trip when off-coil temperatures fall below 2°C, disabling the supply air fan. Additionally, when the outdoor air temperature falls below 2°C Stage 1 frost protection should start. This involves enabling all the heating system pumps and fully opening control valves on non-operating plant until outdoor temperatures rise above 2°C once again.

P3: Dry Workplaces During Winter

Evidence from BUS Methodology occupant satisfaction surveys suggest that it is becoming more popular for occupants to perceive indoor air as dry [62]. The thermodynamic properties of air are defined by the psychrometric chart, show in Figure 5.5 [63]. When air undergoes a sensible heating process, it's dry bulb temperature increases whilst it's moisture content remains similar. As warmer air is capable of vapourising a greater quantity of water with higher enthalpy, sensible heating processes lower relative humidity. Within temperate regions, it is typical for cold air to undergo a sensible heating process before being circulated within a closed door environment. This can often result in indoor relative humidity levels falling to 20%. Influenza virus transmission studies show that spreading of repiratory diseases is most efficient at relative humidities between 20% and 35%. Therefore, this practice is contributing to a workplace health challenge.

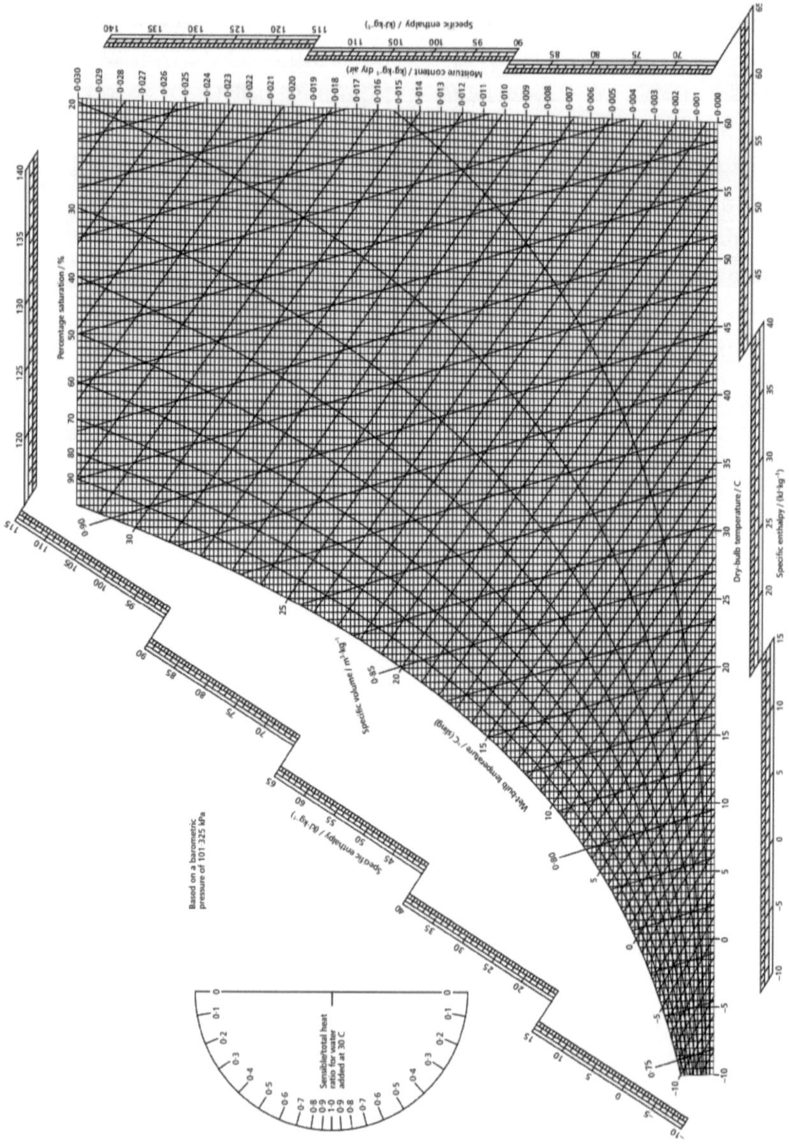

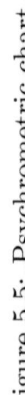

Figure 5.5: Psychrometric chart.

It is recommended that indoor environments be maintained at 40%-60% relative humidity for optimal performance. To achieve this it may be necessary to introduce humidifiers to conditioned fresh air. However, there is a significant risk in using humidifiers which rely on aerosols. Legionnaires' disease is a type of pneumonia that relate to man-made water systems where water temperatures are warm and fairly static. The disease causing bacteria is also encouraged by nutrients in water, limescale and a biological sludge, biofilm. Any humidifier that produces an aerosol is a cause for concern. To address these risks supply chains need to:

- avoid risk of stagnant water;

- avoid allowing cold water temperatures to rise between 20°C and 45°C;

- use stainless steel or food-grade plastic pipework and fittings;

- use effective water treatment, including water softeners and reverse osmosis;

- use the correct humidifier for the expected occupants, selecting less risky humidifiers for high-risk groups.

P3: Wayfinding within Workplaces

The commercial nature of property development means that indoor environments are becoming ever less distinguishable. It is now common to arrive on the floor of an office or block of flats and be entirely disorientated, even when visiting frequently. With the uniform aesthetic and equal treatment perhaps in favour, the role of signalling and signage is important. Evidence from BUS Methodology occupant satisfaction studies suggest that this issue may be most pressing with hall call lifts. Often, when entering lift halls, occupants frequently miss noting their designated lift when they have the opportunity. Just a casual conversation or glance at a phone at this touch point is an understandable reason to miss this information. A better service may be achieved by employing porters to assist occupants and visitors with hall call lift systems.

P3: Skills Gaps

This universal agenda is a focussed response to issues where much demonstration has been observed. The actioning of these items constitutes a substantial inter-displinary effort that requires deep understanding of both the social and physical sciences, including how they interact. Further, it is understood that this agenda is incomplete and may be added to through further experience and perspective. To achieve any of the actions itemised here, there is a need to engage with an audience from what is admittedly a niche starting position. Through such engagement, it is hoped that an industry can be inspired to learn skills where there may be sustained demand. It is hoped that in publication of a positive dialogue anyone may constructively contribute and transform what they can.

P4: Appropriate Real Estate Performance Benchmarking

Historically, real estate has struggled to benchmark relative performance against peers. Every asset may be considered relatively unique and any financial gains can often be the result of performance improvements made by others. Real estate markets are characterised by monopolistic competition and free-riding. Therefore, there is a need to establish a simple common-sense benchmark that signals contributions by landlords to tenants and the wider community.

This can be easily achieved by scrutinising landlords service charge. The following ratio indicates the relative performance of a landlord in management of a real estate asset.

$$\frac{\lambda_\tau \cdot C_\tau}{service\ charge} \quad (5.1)$$

Where, λ is the *Commonwealth Cost of Carbon* and C_τ is *greenhouse gas emissions of service calculated from source to end-use*, for time-period τ.

5.2 Association

This document constitutes a Public commitment of the authors to governance of a platform ecosystem which is the Property of REALFEED Ltd. REALFEED is a company limited by shares registered with Companies House, a government agency of the United Kingdom. The purpose of REALFEED is to act as an agent for one person. REALFEED works within a Society of Peoples, committed to the following principles [64] [65]:

- Peoples are free and independent, and their freedom and independence are to be respected by other Peoples;

- Peoples are to observe treaties and undertakings;

- Peoples are equal and are parties to the agreements that bind them;

- Peoples are to observe a duty of non-intervention;

- Peoples have the right of self-defense but no right to instigate war for reasons other than self-defense;

- Peoples are to honour human rights;

- Peoples are to observe certain specified restrictions in the conduct of war;

- Peoples have a duty to assist other Peoples living under unfavourable conditions that prevent their having a just or decent political and social regime.

5.3 Licensing

REALFEED has no interest in monopoly power over its markets and is keen to learn as much as possible from other agents and organisations. The agency is keen to stay agile whilst commanding a stable knowledge base, so that it can adapt whenever necessary. REALFEED tends to favour permissive licenses for its Property, such as the MIT license applied to this document. It is understood that these arrangements should support conditions of fair liberty and equality of opportunity.

5.4 Contracts

REALFEED is not currently open for appointment of Professional services directly. It is an agency that intends to share knowledge in concert with other Professionals to develop mutual assured approaches for their Clients. Any popular attention paid to REALFEED would be purely a result of the companies Good faith.

CHAPTER 6

Conclusion

It has been recognised that the prevailing discussion on the governance of platform ecosystems may have contributed to global ecosystem instability in alarming ways. The prioritisation of network effects to deliver a popular market share of consumers tends to be a common Feature. Therefore, this document has sought to diagnose causes of instability and define new platform ecosystem boundaries for more stable outcomes. It is understood that this has been achieved through an investigation of ethics, continuous integration processes and operations.

This investigation appreciates the differences between the *states* of Locke and *State* of Hobbes. The Sovereignty of states and their role in forming civil society are recognised and this platform submits to the necessary assurance regime of a state. However, a significant difference of this platform ecosystem to others is the acknowledgement that a society need submit to a global Sovereign of State which accounts for the condition of the Earth's ecosystem. What follows is the use of a Hobbesian Commonwealth Cost of Carbon as an overall quality of Life indicator and a Rawlsian process of alarm management. The resulting *Commonwealth of Peoples* could adopt traditional principles of justice amongst free and democratic Peoples without conflict.

Complimentors are encouraged to develop their own Features for integration with this ethical Framework. A minimal set of criteria has been identified for a Publicly addressable alarm management system that may allow complementors to develop new Features for end-consumers. It is hoped that these minimal criteria prove a stable enough starting point to generate interest and build a supporting community.

In operation, this platform ecosystem acts as an agent for one person. It's intentions are to share knowledge in concert with other Professionals to develop mutually assured approaches in Good faith. Evaluations of practice have identified a prioritised agenda of platform alarms with universal relevance, each of which have associated recommendations for action. Licensing of Property is highly permissive to support conditions of fair liberty and equality of opportunity. A Public commitment by the authors to this platform has now been made Public and is fully auditable. Our knowledge exchange and learning process through failure continues.

Bibliography

[1] Anka P. Thibaut G. Francois C. Revaux-J. (1968) *My Way* Burbank, CA: Reprise Records

[2] Cusumano M.A. Gawer A. Yoffie D.B. (2019) *The Business of Platforms* New York, NY: HarperCollins Publishers

[3] Locke J. (1967) *Two Treatises of Government* Cambridge: Cambridge University Press

[4] Nozick R. (1974) *Anarchy, State and Utopia* New York, NY: Basic Books

[5] Hobbes T. (1668) *Leviathan*, Oxford: Oxford University Press

[6] Rawls J. (1999) *The Law of Peoples*, Harvard: Havard University Press

[7] National Geographic. (2020) *The Carbon Cycle*, Accessed 26/12/2020, available online: https://www.nationalgeographic.org/encyclopedia/carbon-cycle/

[8] British Antarctic Survey (2010) *Ice Cores and Climate Change, Science Briefing*, Swindon: British Antarctic Survey

[9] Tyndall J. (1861) *On the Absorption and Radiation of Heat by Gases and Vapours, and on the Physical Connexion of Radiation, Absorption, and Conduction. - The Bakerian Lecture*, Proceedings of the Royal Society of London; Philosophical Transactions of the Royal Society, 151, 1-36

[10] Harvey D.L.D. (1993) *A Guide to Global Warming Potentials*, Energy Policy, 21(1), pp. 24-34

[11] Hansen J. et al. (2006) *Global Temperature Change*, Proceedings of the National Academy, 103, pp. 14,288-14,293

[12] Cole S., McCarthy L. (2012) *NASA Finds 2011 Ninth-Warmest Year on Record*, Accessed 30/11/2012, available online: http://www.nasa.gov/topics/earth/features/2011-temps.html

[13] Stern N.H. (2006) *The Stern Review of the Economics of Climate Change*, Cambridge: Cambridge University Press

[14] Rawls J. (1971) *A Theory of Justice* Cambridge, MA and London: Harvard University Press

[15] Coase R.H. (1960) *The Problem of Social Cost.*, The Journal of Law and Economics, 3, pp. 1-44

[16] Sidgwick H. (1877) *The Methods of Ethics* Cambridge: Cambridge University Press

[17] Arrow K.J. (1963) *Social Choice and Individual Values*, New York, NY: John Wiley

[18] May K. (1952) *A Set of Necessary and Sufficient Conditions for Simple Majority Decisions*, Econometrica, 20(4), pp. 680-684

[19] Sen A. (1970) *Collective Choice and Social Welfare* San Francisco: Hoden Day

[20] Waldron J. (1984) *Theories of Rights* Oxford: Oxford University Press

[21] Dworkinn R. (1978) *Taking Rights Seriously* London: Duckworth

[22] The Nobel Prize (2021) *William D. Nordhaus, Facts,* Accessed 04/02/2021, available online: https://www.nobelprize.org/prizes/economic-sciences/2018/nordhaus/facts/

[23] Press Association (2007) *Peerage for climate change economist,* Accessed 04/02/2021, available online: https://www.theguardian.com/environment/2007/oct/19/climatechange

[24] Jacobs-S. Rogers T.N. (2019) *Just 26 of the world's richest men have more combined wealth than the*

poorest 3.8 billion people, Accessed 15/03/2021, available online: https://www.businessinsider.com/worlds-richest-billionaires-net-worth-2017-6?r=US&IR=T

[25] Ramsey F.P. (1928) *A Mathematical Theory of Saving.* Economic Journal, 38(152) pp. 543-559

[26] Cline W.R. (1992) *The Economics of Global Warming*, Washington D.C.: Institute for International Economics

[27] Nordhaus W.D. (1994) *Managing the Global Commons: The Economics of Climate Change*, Cambridge, MA: MIT Press

[28] Hope C., Anderson J., Wenman P. (1993) *Policy Analysis of the Greenhouse Effect. An Application of the PAGE Model*, Energy Policy, 21, pp. 327-338

[29] Tol R.S.J. (1997) *On the Optimal Control of Carbon Dioxide Emissions: An Application of FUND*, Environmental Modelling and Assessment, 2, pp151-163

[30] Interagency Working Group on Social Cost of Carbon, United States Government (2013) *Technical Support Document: - Technical Update of the Social Cost of Carbon regulatory Impact Analysis - Under Executive Order 12866*, Washington D.C.: United States Government

[31] Sen A. (1961) *On Optimizing the Rate of Saving* Economic Journal, 71

BIBLIOGRAPHY

[32] Marglin S.A. (1961) *The Social Rate of Discount and The Optimal Rate of Investment* The Quaterly Review of Economics, 77(1), pp. 95-111

[33] Hayek F. (1944) *The Road to Serfdom* Chicago: University of Chicago Press

[34] European Commission (2020) *World military expenditure and weapons trade*, Accessed 26/12/2020, available online: `https://knowledge4policy.ec.europa.eu/foresight/topic/changing-security-paradigm/world-military-expenditure`

[35] co2.earth (2020) *co2.earth Are we stabilizing yet?*, Accessed 26/12/2020, available online: `https://www.co2.earth/global-co2-emissions`

[36] Organisation for Economic Co-operation and Development (2021) *stats.oecd.org*, Accessed 30/07/2021, available online: `https://stats.oecd.org/Index.aspx?DataSetCode=SNA_TABLE11`

[37] Kalder N. (1955) *An Expenditure Tax*, London: George Allen and Unwin

[38] Prichard H.A. (1955) *Moral Obligation*, Oxford: The Clarendon Press

[39] Arrow K.J. Solow R.M. Portney P. Leamer E. Radner R.; Schuman H. (1993) *Report of NOAA Panel on Contingent Valuation.*, Federal Register, pp. 4601-46014

[40] Dasgupta P. (2001) *Valuing Goods.*, In: Human Well-Being and the Natural Environment. Oxford: Oxford University Press, pp. 122-138

[41] Dasgupta P. (2015) *Disregarded Capitals: What National Accounting Ignores.*, Accounting and Business Research, 45(4), pp. 122-138

[42] Morgan G.A. (1941) *What Nietzsche Means.* Cambridge, MA: Harvard University Press

[43] Deering S. Nokia H.R. (1998) *Internet protocol, Version 6 (IPv6) Specification* The Internet Society https://datatracker.ietf.org/doc/html/rfc2460

[44] IEEE Standards Association (2010) *802.1X-2020 - IEEE Standard for Local and Metropolitan Area Networks–Port-Based Network Access Control* IEEE Standards Association https://ieeexplore.ieee.org/document/9018454

[45] IEEE Standards Association (2015) *X.509 Check: A Tool to Check the Safety and Security of Digital Certificates* IEEE Standards Association https://ieeexplore.ieee.org/document/7428340

[46] OASIS Standard (2019) *MQTT Version 5* OASIS Message Queuing Telemetry Transport https://docs.oasis-open.org/mqtt/mqtt/v5.0/mqtt-v5.0.pdf

[47] Open Networking Foundation (2014) *OpenFlow* Open Networking Foundation https:

//opennetworking.org/wp-content/uploads/
2014/10/openflow-switch-v1.5.1.pdf

[48] Linux Foundation (2023) *What is Open vSwitch* Linux Foundation `https://docs.openvswitch.org/en/latest/intro/what-is-ovs/`

[49] FaucetSDN Device Automated Qualifaction Community (2022) *faucetSDN/daq* faucetsdn `https://github.com/faucetsdn/daq`

[50] nmap.org (2023) *Nmap Reference Guide* nmap.org `https://nmap.org/book/man.html`

[51] Chatnzis I (2023) *Ncrack Reference Guide* nmap.org `https://nmap.org/ncrack/man.html`

[52] git-scm Community (2022) *git* git `https://git-scm.com/`

[53] International Standards Organisation (2020) *ISO/IEC 18033-3:2010 Infomration technology - Security techniques - Encryption algorithms - Part 3: Block ciphers* Geneva: International Standards Organisation `https://www.iso.org/standard/54531.html`

[54] Dierks T. Rescorla E. (2008) *The Transport Layer Security (TLS) Protocol Version 1.2* Internet Engineering Taskforce: Network Working Group `https://www.iso.org/standard/54531.html`

[55] Jenkins Infrastructure Community (2022) *ci.jenkins.io* ci.jenkins.io `https://www.jenkins.io/`

[56] International Standards Organisation (2020) *Travis CI* Leverkusen: Travis CI, GmbH https://www.travis-ci.com/

[57] OpenJS Foundation (2022) *node.js* OpenJS Foundation https://nodejs.org/en/

[58] Flutter (2022) *Flutter* Mountain View, CA: Google https://flutter.dev/

[59] Pallets Community (2022) *Flask web development, one drop at a time* palletsprojects https://flask.palletsprojects.com

[60] Parkinson Aidan T. (2024) *Digital Assurance* Wedmore, Somerset: Realfeed Ltd. https://github.com/aidan-parkinson/digital-assurance/blob/main/digitalAssurance.xml

[61] National Grid (2023) *Carbon Intensity API*, National Grid. https://www.carbonintensity.org.uk/

[62] Arup (2023) *BUS Methodology*, London: Arup.

[63] Farrell M. Race G.L. (2012) *Practical Psychrometry*, London: CIBSE.

[64] Brierly J.L. (1963) *The Law of Nations: An Introduction to the Law of Peace*, Oxford: Clarendon Press.

[65] Nardin T. (1983) *Law, Morality, and the Relations of States*, Princeton: Princeton University Press.

www.ingramcontent.com/pod-product-compliance
Lightning Source LLC
Chambersburg PA
CBHW041807160426
43209CB00015B/1717